AMAZON FIRE TV STICK 4K 2022 BEGINNERS GUIDE

An Easy Guide to Using the Fire Stick 4k Max Device with Alexa: Including Troubleshooting Hacks, Tips and Tricks.

By
Natasha Waku

Table of content

INTRODUCTION 6

CHAPTER ONE: SETUP 9

    Banner features for the Fire TV stick 13

    Installing Apps for Fire Stick 15

    How to download apps using the search function 17

    How to install apps using Alexa 17

    HOW TO MANAGE APPS 18

        How to uninstall an app 18

        How to clear cache 19

    How to hide uninstalled apps 21

    Use the Watch list 21

    How to Add Movies or TV Shows to Your Watch List 22

    How to turn off auto-play for trailers 23

    Alexa Assistant In Smart Home 25

    How To Connect Alexa to Fire Tv Stick . 26

    Using the fire TV with an echo speaker . 27

    Show Me Mode 28

    How to Connect One Fire Tv Stick to Multiple TVs 29

    Can t0 Use the Firestick on A Monitor ... 30

Is There a Monthly Fee for The Fire Stick?...31

How to Set Up Parental Control 33

How to enable parental control on a firestick 34

Change parental control pin 35

Adjust parental controls.........................36

Block explicit songs 37

How to setup Amazon free time on a firestick 38

CHAPTER 2 39

Some Of the General Commands for Alexa 39

Basic and general commands 39

General information...............................40

Set alarms, reminders, and calendars...41

Shopping with Alexa..............................41

General information questions and calculations..42

Talk to Alexa and take advantage of its features 43

Find TV shows and movies 44

How to enable 4K playback...................44

The Best Apps for Fire Tv Stick.............46

How to connect a fire TV stick to a different TV...55

The best setting for your fire stick during setup..56

CHAPTER 3..62

How to change location on your Fire Stick 62

How to cancel prime video channels subscription 64

How to cancel streaming service app subscriptions 65

How to connect your fire TV stick to a public network 67

How to turn subtitles off and on Fire TV setting..67

How to watch YouTube 71

How to watch YouTube on Fire Stick using a browser 75

How to watch HBO Max 75

How to use Bluetooth on your fire TV....77

CHAPTER 4..79

How to play games................................79

How to download games.......................80

Top 6 games for firestick 81

How to use USB OTG on fire TV...........86

How to turn off the targeted advertisement...87

FIRE STICK TIPS AND TRICKS...........88

Fire TV stick FAQs 95

CHAPTER 5: 100

Troubleshooting...................................100

INTRODUCTION

Gone are the days when video rentals and cinemas were the only means of catching exciting movies. In the current dispensation, you can stream movies and TV shows while setting on your couch. Tons of video services have replaced the traditional DVD rental routine, allowing for ease and convenience. YouTube, Netflix, Amazon Prime, Hulu and other online video streaming services can be easily accessed, as long as you have internet connectivity.

The Amazon Fire TV Stick is one of the leading streaming devices on the market. There are a lot of reasons to get a streaming device. For one, you may not have a smart TV that allows direct internet connection, or you may have a smart TV with slow internet connectivity.

The Fire Stick streaming device allows you to connect your regular TV to the internet, effectively transforming it into a smart TV that can stream all your favorite movies and TV shows.

Shaped like a large USB stick, the device comes with a cord that's connected to the HDMI inlet at the back of your TV. When connected and turned on, it lets you connected to Wi-Fi, and you can stream shows on all video streaming services.

Made by Amazon, it comes with an integration of some of Amazon exciting services like Prime Videos and built-in Alexa functions.

This little device is packed with a lot of functionalities and special features. Therefore, this guidebook has been written to help beginner learn how to effectively maximize the full potential of the device.

Use this manual as a reference guide as it contains detailed step by step directions, tips and tricks, troubleshooting hacks and overall guidelines for using the Amazon Fire TV Stick device.

Let's get started.

CHAPTER ONE: SETUP

To connect the Fire TV Stick to your television there are two important connectors on your device. First, you have a micro-USB port for power to go through, which you are going to plug in with a cable that comes in the box.

And then you have the HDMI connector that you will connect to the television or the screen you will be using.

To connect the fire tv stick to your TV, connect the power cable to the micro-USB, then plug it into the TV's HDMI slot.

- Your device will first scan its internal storage to organize it and optimize the pre-installed apps when you use it for the first time. This process can take up to ten minutes, although it will usually be a little less.

- Press the play button on your controller to continue, and start the initial setup process. The first thing you should do is to choose the language you prefer. Scroll through the options and select the language you prefer.

- The next step is to choose and configure your home Wi-Fi network. Select the Wi-Fi from the list, by pressing your remote's middle button.
- Then, type in your Wi-Fi password and your Fire Stick device will connect to it.

- The next screen will be an Amazon sign up page. To use an already created Amazon account, choose the **Register** option, which acts as a login button, and to register a new account, choose **Create an account.**

- The fire TV stick will automatically detect your account once you type in the email and password and press log in. Tap **"Yes, continue as"** to confirm that's the account you want to use.

- To save your password, you can accept the pop up asking you to save it. This means that the Wi-Fi password will be saved to your general Amazon web profile, so you can log in easily with any other device you use your account on, such as Amazon Fire tablets.

- Next screen will be the option to activate parental control. If you go ahead to **activate**, it will need a PIN that will be required each time you go to watch Amazon videos or purchase videos, apps, or in-app items. You can skip this if you don't have young children in your home at the moment.
- And that's it, you have finished the initial configuration. You will access the main menu after a short introduction video. It will show pre-installed or purchased applications, and all the contents you have access to.

Banner features for the Fire TV stick

- **PROCESSOR** ---- MediaTek MT8695D Quad-Core 1.7Ghz GPU IMG GE8300
- **RAM** ---- 1GB
- **Storage** ---- 8GB
- **Video** ---- Up to Full HD 1080p, 60fps HDR, HDR10+, and HLG
- **Audio** ---- Dolby Audio, 5.1 surround sound, 2ch stereo and audio pass-through with HDMI (7.1)
- **Connections** ---- micro-USB output and HDMI
- **Connectivity** ---- Bluetooth 5.0, Wi-Fi 802.11 a/b/g/n/ac, 2x2 MIMO, Remote control with volume and on/off (AAA batteries)

- **Operating system** ---- OS 7, based on Android
- **Dimensions and weight** ---- 86 x 30 x 13mm, 32g
- **Ports** ---- HDMI output, micro-USB for charging only, Bluetooth 4.1
- **Control** ---- remote control, and 2 AAA batteries
- **Included accessories** ---- Remote control, HD extension cable, two AAA batteries, and power cord
- **Power input** ---- 100 ~ 240V 50 / 60Hz input, 5.2V, 2.1A output

Installing Apps for Fire Stick

There is no app store on the Fire TV Stick. All the available applications will appear on its main menu. If you need any other app that is not already available, you would have to download it manually.

To find and download new apps to your Fire TV Stick using your remote follow these instructions:

- First check if your Wi-Fi connection is secured before you begin.
- Press on the **"Up"** button on your remote from the home screen. It will lead you to the menu settings.

- Press the **"Right"** button and select **"Applications"** when you see it.
- Press the **"Down"** button and it will open the application tab.
- A menu of featured games and apps will appear. Select the application by pressing the center button.
- Click **"Get"** to install. Most Firestick apps are free. If they however have a price tag, and you can purchase them go ahead and buy it.
- Once you purchase or download any app, it will show on your home screen. To run the application, click on i5 and obey the set-up instructions if it has any.

How to download apps using the search function

The easiest and fastest way to find apps is by searching for them. To do this;

- Select the search button
- Type in the name of the app you want.
- Select the app from the result list.
- Press the download option to get the app.

How to install apps using Alexa

To download app on Fire Stick with Alexa follow these steps

- Press the **voice button** on your fire TV stick remote to activate Alexa.
- Say the name of the app.

- Immediately the app shows on the screen, download it by saying **"Get"**.

HOW TO MANAGE APPS

How to uninstall an app

After using a streaming service /app for a while on the fire TV stick, you may want to unsubscribe. Fortunately, uninstalling an app from your device is as simple as installing it.

Please note that the Fire TV Stick prevents you from uninstalling pre-installed apps like Prime Video, Amazon Music, Alexa Shopping, and the App Store. You can uninstall other installed apps you downloaded without any problem. And also, if you reinstall the application, it will ask you to re-login.

- Go to Settings > Applications >
- Choose "Manage Installed Applications". It will show all your installed applications.
- To uninstall the application, select it. It will move to the application information.
- Choose the "Uninstall" option. Once done the app will be uninstalled and its data will also be removed from the Fire TV Stick device.

How to clear cache

Amazon Firestick is the popular streaming device used to watch movies, TV shows, live TV, and sporting events on your TV with the help of a strong WiFi connection. From the App Store, you can download different apps to view different kinds of media content. The downside of using apps to stream is that each app will store temporary files on the Fire tv stick. This will end in the app crashing, operating slowly, or lagging. To sort this issue, just delete the application's temporary files.

- From your Settings menu.
- Click Applications
- Now, select "Manage installed apps"
- Choose the application you want to clear.
- Select Clear Cache to erase all junk files.

Note that on your Fire stick you will have to delete cache for each application one after the other.

How to hide uninstalled apps

There are countless times when you intend to keep some apps private from prying eyes. You can hide your installed / pre-installed apps with these steps mentioned below;

1. First click on the app you want to hide
2. From the option listed, click on **hide.**
3. To access the app, you can search for it manually or by using voice search.

Use the Watch list

To use your watchlist on the Fire TV Stick, here are the steps to follow;

1. Open the Start menu of your Fire Stick.

2. Next, find the Display List tab. Select it.
3. Every content on the display list will open. If you are a new user, it will be empty.

4. Press the menu button of the device remote and click the remove from watch list option if you want to remove any content.

How to Add Movies or TV Shows to Your Watch List

Your watch list will be obviously empty as a new user of the Amazon Fire Stick.

To add shows or movies to your watch list, do these steps:

- Go to Videos from the device home screen.
- Select the menu button when you find the tv show or movie you want.
1. Then click the "add to Watch List" option.

You can also search for the movie or show through the search box and follow these same steps.

How to turn off auto-play for trailers

Netflix includes two types of auto-plays. One plays a few seconds after you finish a series. The second plays While you are browsing your Netflix catalog; it pops up a trailer of a movie or series.

Fortunately, now Netflix has added the option to disable auto-play when browsing the catalog. It is a universal option, so it applies to all platforms you use Netflix on, including mobile applications.

These changes can be made from your account settings, but you can quickly access them from the mobile application.

- From your More tab, click on account.
- Select the playback settings options.
- Uncheck "Auto-play trailers"
- Click Save to save your settings.

Alexa Assistant In Smart Home

When you start using the different voice assistants regularly, you realize the full potential they offer. Still, sometimes it takes a little help and inspiration to figure out all the things they can do. Among all of them, one of the most useful that you can take advantage of is controlling your TV with a voice command. We can confirm that there is nothing better than being able to say "Alexa, turn on the TV" without searching for your remote.

How To Connect Alexa to Fire Tv Stick

- Open the Alexa application
- Select the Settings option.
- Now select the TV and Video option.
- Select your device by clicking on its name.
- Then tap Pair my Alexa device.

To complete the process, follow the pop-up instructions as they appear and select which device that you want to control. Once the process is complete, you can start controlling your Fire TV and television with your voice.

Using the fire TV with an echo speaker

- Open the Alexa app
- Go to Devices
- Select Plus and scroll to the bottom of the screen
- Select Combine Speakers
- Select the Home Theater option.
- Select your Fire TV stick device.
- Select up to two compatible Echo devices to use as speakers.
- Select the left and right speakers, if you are using two echo speakers.
- Name your home theater. Please note that the speaker group devices must be in the same room as the TV.
- Follow the instructions on the TV to finish the setup

Show Me Mode

Another big update that came to the Fire TV stick is that you can make use of the TV screen to enhance Alexa results. You can ask things like "Alexa, show me the weather", "Alexa, show me the daycare camera", "Alexa, show me my shopping list" and you will see visual answers.

- Swipe down from the home screen.
- Then tap the Show Mode option.
- Another way is to just say, "Alexa, switch to Show Mode."

To disable show me mode:

- Swipe down from your home screen
- Then turn off Show Mode

- Or you can say "Alexa, exit Show Mode.

How to Connect One Fire Tv Stick to Multiple TVs

It is possible to use only one Fire Stick on one tv, if you want to make use it on another tv, you have to remove it from the current TV and plug it into the HDMI port of the new TV.

You can achieve this however with an HDMI splitter. You will have to plug the Fire Stick into the HDMI splitter; it will then transmit the content of the Fire Stick to the HDMI outputs the other TVs are connected to.

Can t0 Use the Firestick on A Monitor

One of the simplest ways to watch movies and shows is using an Amazon Fire TV Stick. All you have to do is plug it to your TV, connect it to Wi-Fi, and start watching. What if you're not sitting on the couch or close to your TV?

Can you link your Fire Stick to your laptop?

The quick answer is yes— but only if the proper equipment is available with you at that moment. Your monitor has to have an HDMI port or a VGA adapter.

If you want to mirror your monitor to the TV instead, follow these steps;

- Turn on the TV and connect your Fire Stick to the port.

- Press and hold the home button of the remote.
- It will pop up a menu, select mirroring from the list.
- Connect your PC to the same wireless network as your Fire Stick.
- After selecting the Fire Stick option, Your PC's screen will now be mirrored on your TV.

Is There a Monthly Fee for The Fire Stick?

You do not have to pay a monthly fee for the Fire TV Stick itself. However, the services you use may require you to pay a monthly or yearly fee. For the Fire TV Stick on its own, you only have to purchase it. Afterward, you would have to pay for the streaming services that you will use.

Amazon may also provide free bundles or subscriptions with some services. Some of the streaming services you would have to pay for are:

- Netflix
- Amazon Prime.
- HBO
- Disney plus.
- Apple TV +.
- Hulu

You will spend a lot on subscriptions if you use numerous streaming services or paid apps. It will be best if you limit your streaming services to a number you can comfortably afford and choose only those that meet your streaming choice.

How to Set Up Parental Control

Children increasingly control new technologies better so they barely need your help to put any type of content on the Smart TV or a device like the Fire TV stick. The main advantage of parental control is that you will not have to be with them all the time to know whether they are not misusing the device. That is, you can limit their use so that they do not access certain channels or certain applications. It is also possible to limit their use to prevent them from buying paid content or applications from your card. You can also decide the kind of series they can watch and also prevent them from seeing explicit photos in
Prime Photos if you have the app on the Fire TV Stick device.

How to enable parental control on a firestick

- Take the Fire TV remote control and open "Settings".
- Select the "Preferences" option.
- Then click on the "Parental Control" option.
- Choose the "Parental Control" option again to open all the functions.
- Set a PIN.
- Then go to the remote control and click the "Menu" button.
- When you have set the PIN, all the parental control parameters available will be displayed. You can then customize them according to your requirements.

Change parental control pin

- Take the remote control and open the Fire TV main menu settings.
- Then go to the "Preferences" option.
- Then select the "Parental Control" option from the list.
- Now, enter the current PIN using the remote control.
- In the new window, select the option "Change the pin".
- After this, a new window will appear. Enter your current login credentials.
- Then, in the next window, enter your new PIN. Enter it again to confirm it's the same.
- Then save it.

However, you can make use of the "www.amazon.com/pin" link to also reset the PIN without using your Fire TV. This link will open the Amazon login page. Enter your existing login details. Then you can set a new PIN.

Adjust parental controls

Once you have an active parental control on your Fire TV Stick, you can adjust the settings and block certain applications or services:

- You can choose to activate whether or not the PIN is required to make purchases on Amazon. This is advisable to do if you have children at home.
- Access to Prime Photos: You can set that you have to enter the PIN

to be able to see all the saved content if you make use of the Amazon photo storage service.
- Age restriction: This option will enable you to restrict content that is not suitable for kids. In this way, nothing that could be for adults will play without the set pin.

Block explicit songs

You can block any song with explicit languages from playing on your firestick device when your kids are watching contents from the fire TV device. However, you cannot activate it from you fire TV as it only works from your Amazon music account:

- Open your Amazon Music app on the fire TV menu.

- Now click on your name in the upper right corner of your screen.
- From the list, click Preferences.
- Next, find the "Block explicit songs" option and check the box next to it.

How to setup Amazon free time on a firestick

For younger children, you can turn your Fire TV into a super kid-friendly walled garden with Amazon Free Time. To do this:

- Go to Settings
- Select Free time and parental controls".
- Then choose "Free Time Settings".

CHAPTER 2

Some Of the General Commands for Alexa

From managing your to-do lists, ordering purchases, asking to play a song, telling it to wake you up in the morning with your favorite song, asking for the closest Chinese restaurant to you, etc. Whatever it is, you can say basic commands to make the most of your Smart speaker.

Basic and general commands

- Alexa, stop
- Alexa, repeat
- Alexa, turn up the volume
- Alexa, turn down the volume

- Alexa, next/previous (in case of songs or news, for example)
- Alexa, stop / mute/stop
- Alexa, restart
- Alexa, read my notifications
- Alexa, delete everything I said today
- Alexa, turn on whisper mode (you can speak lower)

General information

- Alexa, what time is it?
- Alexa, what time is it in New York? Let's you know the time of any location you say.
- Alexa, what's the weather?
- Alexa, what's the weather like in Madrid?
- Alexa, is it going to rain today?
- Alexa, how windy is it?

Set alarms, reminders, and calendars

- Alexa set an alarm
- Alexa set an alarm for 7 AM
- Alexa, wake me up at 8wm with music from Queen
- Alexa, cancel all my alarms
- Alexa, what alarms do I have?
- Alexa, snooze
- Alexa set a reminder
- Alexa, delete the reminder
- Alexa set a timer for 5 minutes
- Alexa, what is on my calendar today?

Shopping with Alexa

- Alexa, I need to buy water
- Alexa, how is my cart

- Alexa, how is my purchase
- Alexa, what restaurants are near me?
- Alexa, find a nearby pharmacy
- Alexa, what are the hours of the Prado Museum?
- Alexa, what movies are playing near me?

General information questions and calculations

- Alexa, when did Edgar Alan Poe die?
- Alexa, what does Everlasting mean?
- Alexa, how do I say 'Thank you very much in Chinese
- Alexa, when did the first man land on the moon?
- Alexa, how far is the Moon?

- Alexa, what time does the sun rise?
- Alexa, what is the square root of 99?

Talk to Alexa and take advantage of its features

- Alexa, what can you do?
- Alexa, surprise me
- Alexa, why is your name Alexa?
- Alexa, tell me something
- Alexa, call XXXX: Replace the XXXX with the name of a contact on your phone who also has an Amazon Echo.
- Alexa, tell me a story.
- Alexa, sing a song.

Find TV shows and movies

With your Fire TV Stick, you have access to an unlimited number of paid or free apps that have thousands of popular tv series, shows and movies. You will find a list of recommended media streaming apps below.

How to enable 4K playback

Follow these steps to enable 4K content on your device:

- Open the Settings app of your Fire Tv Stick
- Open the "Screen and Sound" menu.
- Click on the "Display" option.
- Press where it says "Video Resolution".

- All the resolutions that are compatible that can play on your TV will be listed. Most likely, the "Automatic" setting will already be enabled here.
- You should choose 4K playback ideally, i.e., "2160p 60 Hz" or "2160p 50 Hz" options.

To know which configurations is better for your TV follow these steps to find out which resolution is the best:

- Go to "Screen and sound".
- Choose the "Audio and Video Diagnostics" option.
- Your fire TV stick device will show you the maximum compatible resolution for your TV and also the video functions.

The Best Apps for Fire Tv Stick

Kodi (Free)

Kodi is an essential application that should not be missed on any device, especially on your TV screen. It has a system of add-ons with which can add different types of functionalities depending on your characteristics and you are free to change the theme to suit you. The app has to be manually downloaded from the official Kodi website.

Netflix (paid)

If you want to enjoy series, movies, and documentaries on your TV, it is clear that Netflix is one of the best alternatives. Whether you have an account for yourself or you share with your friends or family, you'll have wide possibilities to enjoy. You also have access to suggestions for recommended content that suits your tastes.

BBC iPlayer (Free)

BBC iPlayer is a popular streaming service that mainly broadcasts British television shows. It however, has a global audience numbering in the millions. Some programs are available outside the UK, but not all. You may have to use a VPN on the app if you reside outside the UK.

Expressvpn

This British Virgin Islands VPN provider offers good download speeds and excellent options for mobile iOS, Android routers, Fire TV, Xbox, PlayStation, Linux, and many more. It supports up to 5 devices and it doesn't keep any logs. It works perfectly for Netflix or Disney Plus. It has a large number of countries to choose from. Their servers are once again very fast, a very viral feature, if you need them for online video games.

Crunchyroll (paid)
If you like anime and manga, then Crunchyroll is one app you should have on your device. The app gives you access to the Crunchyroll streaming service, with over 25,000 episodes translated into multiple languages. Crunchyroll is a premium app. It does not offer a free service, although if you understand and get used to the interface then you can register for the two weeks free trial before you decide to pay for the subscription.

In its Premium version, you have almost the same broadcasts as those in Japan, good HD quality without ads and to the complete catalog. With the free version, you will have access to only a small part of their catalog, with poor image quality and lots of ads.

Disney plus (paid)
The long wait is over. Disney has officially launched its Disney Plus app for Amazon Fire devices. The app was released in November 2019 and it can be downloaded from the Amazon app store. It is a subscription streaming platform for you to watch all your favorite movies and TV shows and allows you to stream 4 devices simultaneously. It also has content for children.

Cinema app (free)

This app came about shortly before the closure of Terrarium TV and became mainstream afterward.

With hundreds of hours of playable content and an endless content lineup, Cinema APK is one of the best entertainment apps for Fire Stick. The app itself gets regular software updates, which gets better over time.

HBO Now (paid)

If you are a fan of HBO content, HBO NOW is a must-have streaming service. You have access instantly to all HBO originals like movies, news, comedy, documentaries, and more. HBO NOW is also home to several award-winning series. It is a subscription-based service and you are able to cancel the subscription anytime you want.

HBO is free for new users as they will get a free trial. Their subscription is available at $14.99/month is the best option so you can stream with no limits.

HBO NOW is currently only accessible in the US and certain UK territories. You would have to use a VPN or web proxy to use HBO NOW on Fire Tv stick if you do not in the specified regions.

CatMouse apk (free)
CatMouse is a free streaming service like Netflix, Hulu, Amazon Prime, and other platforms. It does not require any subscription for you to stream on the app or have it on your fire TV device. It

was made to help users enjoy premium media entertainment without having to spend money.

Typhoon TV (free)

Typhoon TV is a free streaming app that gives you access to stream popular series, movies and TV shows. It is the clone of the old and popular Terrarium TV. The app also has support to download and stream your favorite media in offline mode. With services like Open Subtitle or Subscene, you can download subtitles for all media content available in this app.

Cyberflix TV (free)

Shutting down Terrarium TV resulted in many clones spawning; CyberFlix TV is one of them. And since the CyberFlix TV is the clone, it looks exactly like

Terrarium TV. The good news is that it also works a lot like the original.

CyberFlix TV includes a very wide collection of shows and movies which it sources from the same place as the Terrarium TV.

Sling TV (paid)
Sling TV was the first live TV launched on the webspace and it should be among your first choices if you like live TV channels. However, their service is available to regions in the US only, you need a VPN to access it elsewhere.

Most live video streaming services charge upwards of $40 for the same features that you can get for just $20 a month with Sling TV. The only concern is that you need a proper internet connection to enjoy uninterrupted live streaming services.

The only restriction is that if you opt for the cheapest package, you can only stream on one device at a time. In case you have signed in with two or more devices, they inquire from you the device you will be using to watch channels with when you start.

Mobdro (Free)

If you like to watch sports and not spend a dime, then Mobdro is the right Fire TV Stick app for you. Do not search for this app on the Amazon App Store because it can only be downloaded from the Mobdro official website. You can install it from Kodi.

Once you download this app and run it on your Fire stick, you will notice that Mobdro continuously searches for free videos you should watch and sends them to you, so it is a very useful tool in this regard.

How to connect a fire TV stick to a different TV

Since the Fire TV Stick device plugs directly into your TV's HDMI port, they can be easily plugged into a new TV. Be sure you have all the necessary accessories with you, including the remote, power cord, HDMI cable, and the Fire TV itself. Follow these steps to connect it to the tv;

- Find a power outlet near the TV.
- Plug the Fire's cable into the wall.

- Plug the other end into your device.
- Connect the fire TV to the tv with your HDMI cable.
- Turn on the TV and find the right input channel.
- Your Fire tv's home screen should be ready for you to start streaming and accept remote input.

The best setting for your fire stick during setup

Here are some fire TV stick settings that will improve your overall experience with the device including your enjoying your privacy and streaming experience.

When you first buy your device, it will come with a pre-installed setting known as factory settings. You can start using your Fire TV stick without changing those settings. Or you can change them to better suit your needs.

1. Enable installation of application from unknown sources and app debugging:

To use your phone or your computer to download apps to your Fire Stick device, there are some steps you'll need to take first. These aren't that different from enabling ADB debugging on your Android smartphone, but it's a bit different since we're doing it from a Fire TV and not a phone.

- Turn on your Amazon Fire TV Stick.
- Open the Settings app.

- Scroll down to My Fire TV.
- Select the Developer Options.
- Click ADB Debugging
- Enable the option "Apps from Unknown Sources"

After you enable ADB debugging and allow apps to be installed from sources other than the Fire TV Store, you'll need one more vital piece of information: the Fire TV's IP address. Here's where you can find it on your device:

- Turn your Fire Stick device on.
- Click on the Settings app to open it.
- Locate Fire TV IP Address.
- Navigate to Device.
- Select About.
- Locate Fire TV IP Address

- Scroll down and choose the Network option.
- Locate Fire Stick IP Address.
- Write down the indicated address.

2. Change App Notification Settings:

The topic of notifications in the world of TV devices depends a bit on taste. For some, you may find notifications as useful as they are on your mobile, while others hardly pay attention to them. If you belong to the last category of people, you may be interested to know that you can disable it, so that they do not interrupt you when you are viewing content.

To do so;
- Go to Settings

- From there go to Preferences.
- And click the Notification Settings option.
- Enable the "Do Not Disturb mode"

Also, from the Application notifications, you can decide which applications can send notifications, or just directly disable all of them from sending notifications.

3. Disable Video and Audio Autoplay

It is very annoying when you are browsing on Netflix, and videos keep popping up at you in the thumbnails of the content. Netflix does not allow you to deactivate but on your Amazon Fire

TV Stick, you can stop video playback and sound automatically. To do this;

- Go to "Settings - Preferences - Sponsored content" and check the "Disable automatic video playback" box.

4. Manage Your Privacy Settings and Turn Off Data Monitoring

The Fire TV device has several options to disable or enable your privacy settings. If you go into Settings, and in the top bar go to Applications. Enable or disable the "Compile app usage data". There is another privacy section in the preferences settings with more options.

Specifically, in the privacy section, it has two key options. You can disable the fire from collecting data based in your usage and you can do the same with the data related to the usage of your installed applications.

CHAPTER 3

How to change location on your Fire Stick

Like every other Amazon device, the fire TV device identifies your exact location and tracks your data. Some people don't know or don't care. While for some, it is a serious issue to worry about.

The situation is that turning of your location is more complicated than turning off your data. This is because the fire stick does not have any option to turn off your location. The only way you do this is to make use of a VPN application.

VPNs are the perfect option to help you to change your location on any device and make it look like you are in a different country entirely.

VPN is the short form of virtual private network. It provides a wall of privacy and security to the Internet users that connects to them.

The first users of VPN is dated back to some companies who use VPNs to preserve their company whilst some of their workers worked remotely.

A VPN is often likened to a blanket. It protects the user from security issues and gives ultimate privacy like the blanket gives warmth from cold. It provides adequate security against hackers and prying eyes from companies.

Finding a VPN for your phone or computer is very easy and you can also easily find a VPN for your Firestick too:

- Go to your device's main menu.
- Click the search icon and type in "VPN" or the name of the vpn you want.
- Then follow the installation guide given.

How to cancel prime video channels subscription

How to cancel prime video channels subscription

- On a computer, launch your browser and visit PrimeVideo.com.
- Select your profile icon
- Select Account and Settings.
- Click the Your Account tab.
- Under the membership header, select End Membership.
- Confirm and cancel your subscription to the video.

How to cancel streaming service app subscriptions

Here's how you can cancel your subscriptions on streaming apps:
- YouTube

Choose an account from your icon and select the plan details option. It will display if you have an active subscription and a link to cancel. To cancel, select the cancel subscription option under the membership and billing section.

- Amazon Prime

To cancel as a monthly subscriber on Amazon prime, go to your account's prime membership and click on update, cancel and more. Follow the pop-up instructions to cancel your Amazon prime subscription.

- HBO Max

Go to HBO's help page and search for the option to cancel your subscriptions. Follow the link and instructions given on the page.

Try to end your subscriptions at least a day or two before you are billed for the next month. If not, your access will be renewed automatically for another month.

How to connect your fire TV stick to a public network

Do you remember that your Fire TV can install browsers? Well, that will be of help when you go to connect to a public network in which you are asked for a password. Take for instance you are at a hotel. From your Network settings, enter the hotel's Wi-Fi, and if you need to enter a password, the hotel website will open in your browser just as it would on your mobile or laptop. Put in the password and now you can make use of the Wi-Fi without any hitch.

How to turn subtitles off and on Fire TV setting

Closed captions are necessary for some viewers to enjoy the content of a program. Whether you are hearing impaired, want to watch a program in a foreign language, or prefer to not have subtitles, turning on/ off subtitles is easy.

Fire TV setting
- Play the video you want to watch.

- Press the menu button on your Fire TV remote
- Click on the "Settings" option.
- Go to "Accessibility".
- Select "Closed Captions" from the list.
- Turn on the "Closed Captions" switch.
- Return to your video by pressing the "Home" button on your Fire TV remote.

To disable them, follow the same steps, but click "Disable".

Amazon prime video

- Play a movie or show you would, want to watch.
- Look for the icon (it looks like a chat box) in the upper right corner and click.

- Choose your preferred subtitles language
- Change the size and color of the font as it suits you.

To disable subtitles, follow the same steps but click the "Off" option instead of choosing a language.

Netflix
- Launch Netflix.
- Choose a movie, TV show, or video.
- Once playback starts, press the down arrow on the remote.
- Select the dialog icon.
- Confirm the selection.
- Make your subtitle changes (turn them off or on).

Hulu

- Navigate to the Hulu website.
- Choose a movie, TV series or show you want to watch.
- Move the cursor over the playback area to bring up the playback controls.
- Select the tools option at the bottom left.
- Click Subtitles and Audio for subtitles and language settings.

Disney plus

- From the Disney plus webpage Click on the subtitle icon in the top right corner of the movie or video you are watching
- Then click on the gear icon to activate/deactivate subtitles

YouTube

- Hover over a YouTube video for the video settings which becomes visible when you pause the video.
- You can enable or disable the subtitles by clicking the red button "CC"

How to watch YouTube

YouTube is the most famous and widely used video-streaming website on the web today. Millions of users have uploaded videos on it and some videos have exceeded one billion views to date. It is part of the few applications that all Internet handlers use in the world. Their service is accessible through a dedicated app on platforms like Android phones/tablets, iOS (iPhone/iPad),

Roku, Amazon Firestick, and more. In 2018, the YouTube app was removed from the Amazon app store after some agreement issues between Google and Amazon.

Customers always want to use YouTube on all their devices and when it is not available on the Amazon app store then it will be the biggest disadvantage of Firestick. However, the issue between them; two digital giants has now been resolved and YouTube can now be downloaded through the Amazon store. Other Google apps like YouTube TV and YouTube Kids are also available in the store.

As mentioned earlier, YouTube is now officially accessible from the store and you can install it on your Fire TV stick easily and quickly.

- Launch the Firestick and click to the search menu.
- Type and search for YouTube.
- YouTube along with its related apps will appear in the search results.
- Select the YouTube app.
- To install click Download or Get.
- Wait while YouTube is downloaded and installed on your Firestick.
- Once installed, click Open to launch the YouTube app. You can also run the app from the Apps section.
- Select SIGN IN if you want to sync your YouTube app across different devices or select Use

YouTube without signing in if you want to use the app without providing your Google account credentials.
- Once the Sign-in option is selected, the activation code will display on your screen.
- Now, visit youtube.com/activate on your mobile phone or PC
- Tao next as soon as you enter the activation code. After activation, the YouTube app on Firestick will automatically refresh to display the videos.

How to watch YouTube on Fire Stick using a browser

- Install any browser (Firefox or Silk) on the Firestick.

- Open the browser and type in the YouTube URL.
- The YouTube page will now open in the browser.
- Login with your Gmail and password. Upon activation, start streaming your YouTube videos.

How to watch HBO Max

You can immediately install HBO Max on your Fire TV Stick by just opening its app store, and searching for the official HBO Max app. Once it shows up in the search results, you can install it like any other app.

To install the HBO Max app on your Fire Stick device then follow these instructions:

- Open the Amazon Fire TV Stick interface and tap on Settings.
- Now, click on My Fire TV
- Select the Developer Options and enable Debugging ADB and Apps of unknown origin. With this, you can now install APK files on your Fire TV Stick.
- Now, head over to the Amazon Fire TV app store and download the Downloader app.
- Once downloaded, go to Categories and select Utilities.
- Within this option, click Browser and search for Aptoide TV.

Aptoide TV is an app that will make things very simple for you. Mainly because you will only have to search for the HBO Max application in the Aptoide search engine and install it on your device.

How to use Bluetooth on your fire TV

It is possible to connect a Bluetooth headset or speaker to your Fire Stick. If you are watching a movie late into night and don't want to disturb other people around you, you can connect your fire stick to a Bluetooth headset. Or, if you're looking for a somewhat cinematic audio experience, you can pair your Bluetooth-enabled external speakers with your Fire TV Sticks.

Follow these steps to do that:

- Turn on your headphones or speakers.

- Turn on Bluetooth and run on the "discoverable" mode.
- Go to your Fire Stick Settings
- Select the option "Bluetooth Drivers & Devices"
- Click "Other Bluetooth Devices"
- Then select the option "Add Bluetooth Devices"
- Wait while Fire Stick searches for available Bluetooth devices
- Once your device shows, click on it to pair it with your Fire Tv Stick

CHAPTER 4

How to play games

Beyond watching series or making use of your Smart TV, you can also download games for Amazon Fire TV Stick without having to have a game console. Although there are some paid ones, most of them are free and will allow you to play games without having your phone or tablet.

There are games for Fire Stick device that require you to have a remote connection to the TV via Bluetooth, but there are also others that you can play directly with the device's remote. When downloading it, on the description screen you will see whether you require a game console or not.

How to download games

- Open Amazon in the browser
- Log in with your username and password
- Open the menu on the left side of the screen
- You will be given some options
- Click the option "Apps and games for fire stick"
- Here you can choose "Games" and in the filters, on the left, you can select your specific model or if you want to filter by language, by price (if they are only free), or by rating.
- On the right side, you will see the linked devices and a drop-down menu with which to choose the one you want. Choose if there are

several available and touch "Get it now in 1-Click".

If it is not compatible with the model you have registered, you will see an "x" appear next to the name of the device.

Top 6 games for firestick

1. Asphalt8: airborne

The Asphalt saga is like the most important in the world of driving games for mobile phones, and the version for tablets is available on the fire stick device. The game will allow you to drive amazing cars in no less attractive locations.

Asphalt8 has excellent graphics and has over 200 bikes and cars for you to select from, with hundreds of options to customize and improve them. The game has dozens of races, with various events taking you to cities around the world. Price: Free

2. Badlands

A game where you have to fly with a creature as if it were the Flappy Bird and overcome all kinds of obstacles. Along the way, your character will multiply and you always have to try to get one of the clones left by overcoming the traps, even if that requires sacrificing some of the ones you carry with you.

Its artistic section is excellent, with careful graphics but not too presumptuous. The game has a single-player campaign with more than 100 levels, a multiplayer campaign with more than 30 levels, with also a cooperative mode for up to four players. Price: Free

3. Beach Buggy Racing

If Asphalt is a fun game that tries to embrace realism and good graphics, Beach Buggy Racing is almost the opposite. It is a racing game in kart format, based on classics such as Nintendo's Mario Kart. The structures are the same; compete and pick up weapons and special advantages at different points on the many maps that you can go through. Price: Free

4. Bomb Squad

A game that you will enjoy better if you connect a peripheral to improve the gaming experience. Your mission will be simple, to eliminate all the rival characters by punching them, making them fall off the platforms, or throwing bombs at them.

It has several game modes, from single-player to cooperative or multiplayer, and the most enjoyable way for you to play is with your friends on the same television. It is based on mini-games, ranging from bomb hockey to slow-motion death match to classics like capture the flag.
Price: Free

5. CrazyTaxi

It is an arcade game that Sega created for arcades and then brought to the Dreamcast. The version for Android that Amazon offers for your device is ported precisely from the one for the console, and the mission is once again to go around the city at full speed driving your taxi like crazy.

During the tour, you will have to collect fees and collect points that add 3, 5, or even 10 minutes to continue playing. Nor should you forget the soundtrack of the game, in which nineties groups such as The Offspring and Bad Religion participate. Price: 3.69 euros

6. Crossy Road

An imperishable classic perfect to burn your free hours on any platform. The game is very simple and stands out for the simplicity of its controls and its graphics made up of large pixels. You start the game with a hen, and your mission is to get as far as you can crossing roads without being run over and jumping lakes between logs.

You will earn a number of coins, as you progress in the game. And with those coins, you can unlock dozens and dozens of different characters to cross the road with. On many occasions, some character types will also change the look of the stage entirely. The mission is simple, always try to get as far as possible and get better. Price: Free

How to use USB OTG on fire TV

With an OTG cable, you can connect your firestick, smartphone, tablet, etc. to an external device. It is an inexpensive tool to upgrade your device. With your OTG cable, you will be able to increase the storage on your firestick device. You will also improve internet connections and upgrade its functions.

To do this, you need a 3.0 USB cord, and an OTG cable.

What you need to do is;

- Connect your OTG to the fire stick
- Power on your TV, before plugging in the USB drive
- A menu will pop up asking you what you want to use the USB to be. Select device storage.

You will have to format your USB drive for it to work on your fire TV stick and unless you format it again, you cannot use it on another device.

How to turn off the targeted advertisement

In the privacy section, you can manage the ads that appear on your screen. While you can't stop them from appearing, you can turn off Interest-Based Ads. This will at least not show the ads that Amazon thinks you might be interested in based on what it knows about you.

These interests are managed by an ad ID, which you can reset if you think Amazon has too much access to things you are probably interested in or not. Go to settings, click on privacy and turn off targeted apps.

FIRE STICK TIPS AND TRICKS

1. **Sideloading apps**

You can install third-party apps outside of the Amazon app store. Go to the system configuration, and in it in Device. There, activate the ADB debugging and Apps of unknown origin options within the Developer options category.

Once it is enabled, you will be allowed to install APK applications that you download from third-party pages. Of course, keep in mind that doing so can be quite dangerous if you do not do it from a site you trust, since external applications will not be subject to Amazon's quality, stability, or security controls, and some may try to infect you with them.

2. Change the name

You can only change the name of your device from your Amazon profile. Therefore, start by logging into Amazon, and clicking on the option Manage content and devices that you have in the menu that appears when you go to the Account and lists section.

In this menu, click on the Devices tab. It will show everything you have linked to your Amazon account. In it, click on your Fire TV Stick and click on the Edit button that will display next to the current name that was assigned to it in the initial configuration.

3. Fire TV remote app

Amazon has created an application which can make your mobile a remote control to operate any Fire TV device. It's known as the Fire TV Remote App, and you can download it for Android on Google Play and iOS on the App Store.

The application, had the same buttons and function as that your controller has, and you will have a trackpad to be able to move through the menus. But in addition, at the top, you will also have shortcuts for functions such as the main menu, settings, device applications, and the mobile keyboard.

4. Disable Auto-Play in video

If you go into Settings, go to Preferences and click on the Featured Content option, you can disable the Allow Auto-play option. Doing so will cause videos to automatically stop playing in the featured content carousel that appears on the main Fire TV screen when you hover over them.

5. Get a workout

You can stay fit even if the fitness classes or gyms around your area are closed. All that is needed is your fire TV stick device and these apps:

1. YouTube

The YouTube app gives you access to a ton of free workout videos. You just need to search for them using the app or the website in a browser and you will have a variety of workout videos for you to choose from.

2. 7-minute workout

It is an application that bases its proposal on high-intensity circuit training or HICT and is very appropriate for those looking for intense training that lasts a short time. The app proposes 12 routines with periods of 30 seconds, after which you have rest periods of 10 seconds.

Therefore, if you don't have the required equipment at home and you can't possibly go to the gym, the exercises in this app won't take you too long.

3. Yoga TV

For something more relaxed, this application will ease you into yoga exercises. It offers three yoga practices for beginners, all of them with videos and instructions to do it correctly. And no, it's not about making impossible poses from the first day, but about getting started and comfortable with the exercises.

- **Repair the remote**

If for some reason your device's remote has stopped working and there is no way for it to link back to your device, there is a button combination that you can use to pair it again. To do so, you just have to press and hold the Home button for ten seconds. The button has the drawing of a house on it.

6. **Restart with the remote**

With no power button, restarting the device from the remote seems impossible. But it is not, so you no longer need to get up to unplug it and plug it back in every so often. There are two ways to do this, one in the system menus and one with a key combination on the remote control itself.

- Go to the Settings menu,
- Select the Device section.

- Then click on the Sleep and Restart options to restart your device.
- You can also hold down the Play and Center buttons for 5 seconds at the same time and it will restart.

Stick it to the click

If you do not like the clicking sound your remote makes when you press a button, here's how to turn it off.

- Go to your setting and click on display and sounds
- Enter your pin
- In the next screen select audio
- Now turn off the navigation sound option

Fire TV stick FAQs

What are the Amazon firestick requirements?

Here is a list of things you need to set up your Amazon fire stick

- An internet connection
- A TV with an HDMI port
- An Amazon account.
- 2 AAA batteries to power the remote

Do you need Amazon Prime for the firestick?

As part of the Amazon ecosystem, when setting up the Fire TV Stick, the installation app insists that you provide your Amazon account information.

This makes it easy to integrate your Fire TV Stick with your Amazon Prime Video account. However, you do not need a prime video for Fire Stick. If you don't have an Amazon account you can sign up immediately.

Is there a limit to how many fire sticks I can have per Amazon account?

You are not limited to a certain number of Fire Stick devices that you can connect to your Amazon account. However, there is a limit on the number of videos that can be streamed at the same time from that account.

Can I set up the fire stick without an Amazon account?

You cannot use your fire stick device without an Amazon account. When you turn on your Fire TV Stick for the first time, it launches to a setup screen. You will be asked to enter Amazon account details to proceed with device activation and gain access to its feature set.

How to connect Amazon Fire Stick to Wi-Fi without a remote?

One of the main requirements to fully enjoy your Smart TV is for it to be connected to a strong Wi-Fi network. Connecting the Amazon Fire Stick to a Wi-Fi network is almost as easy as connecting to a computer or cell phone. However, in this case, you must make use of an extra tool; A remote. Here are some alternatives when you do not have your Fire Stick Remote.

- With the HDMI-CEC remote control

This is a remote that has buttons dedicated to scrolling the screen and indicating which direction to scroll. Using these buttons, you can open the main menu and click on Settings.

When the options are displayed, tap on Networks to have the Fire TV start searching for nearby networks. When you find yours, select it, enter the password, and press Connect. Once you do this, the network should tell you that you are Connected.

- Using a mobile device

As you already know, the Amazon Fire tv Stick application works as a remote control, so once you have it installed and affiliated with your Fire Smart TV, you can use it to connect your Fire TV stick to a Wi-Fi network.

- Buy or borrow a remote

Another possible solution when you do not have remote control and prefer not to use the one emulated by the Amazon Fire Stick application is to buy a new one at one of the Amazon stores or borrow one from a friend or family member who uses Fire Stick and then you go ahead connect your Fire Stick.

CHAPTER 5:

Troubleshooting

Restart your firestick

1. Restart via settings

This is a way to reboot your device when everything is working fine. You may also use it if your Fire TV device freezes intermittently. If your device is completely frozen, then continue on to the next two methods. However, here's how to reset the Fire TV Stick with your remote:

- Go to the Fire Stick home screen.
- Select settings.
- Then select My Fire TV.
- Click on the Restart at the bottom of the list.
- A message will appear, read it and click on the restart option.
- This will start the reboot process.

2. Restart with remote

If your screen is completely frozen and the TV isn't responding to any commands, then you can restart your device with your remote. It will close all running apps. All you have to do is Press and hold the Play/Pause buttons simultaneously for about five seconds. The reboot will start immediately.

3. Unplug your device

This should be your last option, because it will force your fire TV device to restart.

Leave it unplugged for at about a minute or two before you plug it back in and turn on the tv.

Fix the black screen

1.Change accessories

The black screen on the Fire TV Stick could be caused by its accessories. Check that the transmission pen is properly connected to the power adapter using a USB cable. You should also make sure that the micro-USB end of the cable is securely attached to the Fire TV stick. Similarly, you can try to use another USB power cable and check if there are different things afterward.

It is recommended to use the original USB accessories (power adapter and cable) included with the device. Third-party or counterfeit accessories may not be able to provide enough power to boot up the Fire TV Stick and keep it powered up for a long time.

2. Internet connection troubleshooting
It is recommended to have at up to 10 Mbps for 1080p and 20 Mbps for 4K streaming. If the internet connection is slower than this, you will experience

constant buffering which will lead to a black screen

3. Restart Fire TV Stick

As mentioned above, the black screen issue could be caused by a minor bug with the streaming device. Simply restart the device to get back to normal.

4. Try another HDMI port

If the Fire TV Stick black screen issue, please the device was successfully installed all the way (not halfway) to the TV's port. Also check if your TV has multiple HDMI ports, and try switching the Fire TV Stick to another port.

5. Use the HDMI extender

It is possible to connect the Fire Stick device directly to your TV, Amazon suggests that you include a HDMI extender. In addition to promoting remote performance, improving Wi-Fi connectivity, the cable will also enable you to position the Fire TV Stick in the HDMI port on your television properly.

6. Check the input source

It is important that your input slot and the HDMI port the device will be connected to match each other.

Make sure that if your TV's active input source is HDMI 2 the fire tv device is connected directly to it. If not, it may cause a black screen.

This is not a rare issue; it I'd item recommended that you verify that the HDMI port and the active input source correspond.

Fix buffering issues

- Check your internet speed

To fix buffering issues check your internet connection speed first. Streaming video requires quite a lot of networks, and if your connection isn't up to the task, your Fire TV device won't be able to keep up.

It will then buffer since it doesn't have enough video stream loaded to keep your video playing.

You can use any internet speed checker you like.

- Adjust your preferences

There are a few tweaks you can make to your Fire TV Stick that can optimize its overall performance, and this can help resolve buffering and freezing issues.

- Select Settings on your Fire TV Stick menu.
- Select Preferences.
- Select Data monitoring and turn it off.
- Output data monitoring.
- Select Notification Settings.
- Select App notifications and turn off all the notifications you don't need.
- Exit notification settings.
- Select featured content.
- Turn off Allow audio Auto-play and Allow video Auto-play.
- Uninstall any app you are no longer using.

It's tempting to install all the cool apps you see on your Fire TV Stick, but it's important to remember that your device is a computer. Just like any other computer, if you load it heavily, it will start to run slowly and develop problems. If you're having performance issues, just uninstall all the apps you're not using.

- Select Settings on your Fire TV Stick menu.
- Select apps.
- Select Manage installed apps.
- Select an app and select the Uninstall option to remove the app.
- Repeat this process for all the apps you don't use.

Fix crashing apps

In the event that your apps are crashing more often than usual, be rest assured that it can be fixed by just clearing the data files and cache of those apps or reinstalling them.

- Click on the Settings option on the main screen.
- Select Apps on the next screen.
- Click on the Manage installed apps option to view your apps.
- Click on the app you're having trouble with.
- Then click on Clear Cache.
- And select the Wipe data option as well.
- Launch the app and if it still causes problems, select Uninstall to remove the app.
- Re-download the app to your Fire Stick from the Amazon Appstore.

Stop the fire stick from reading your screen

Your Fire Stick comes with a feature where it reads everything on your out loud. Do not fret as this is a preinstalled setting that can be disabled.
- Go to settings and choose the Accessibility option.
- Click on VoiceView.
- Press it to disable the VoiceView option.

Fix Internet issues

If you're having trouble with your internet connection on your Fire TV Stick, try turning off and reconnecting to the Wi-Fi network as follows.
- Go to the Settings menu and select network.

- Click on your Wi-Fi network to disconnect it.
- Wait for a few seconds, then press on it again to reconnect.

Update your firestick

Like every computer or smartphone, your Amazon Fire Stick needs to be updated as soon as a new update is made available. To update it,

- Go to your settings menu, choose the My Fire TV option.
- Click About on the next screen.
- Then select Install Update (if an update is available to install.)

Reset your Amazon fire stick

- From the main screen, click on the Settings option that appears at the top of the screen.
- Select the My Fire TV option.
- Among the options that you will find, are restart and reset. Try restarting before deciding to reset your device. However, if you are certain of your decision, select and click on the option "Reset"
- For security and to avoid errors, a window will appear asking for your confirmation to reset. It will inform you of everything your device will lose access to, including your login and personal preferences. Click on the Reset button to confirm that you want to do it.

And that's it. Now a screen will appear telling you that the device is being reset to default settings. You must not turn off the Fire TV Stick while performing this operation so that you do not encounter any type of error.

Made in the USA
Columbia, SC
14 July 2024

38618587R00065